The Venusian Challenge

The Adventures of Joaquim Vol.2

illustration: Blackbeard

Giuseppe Lobato

Dedicated to my wife and son.

Giuseppe Lobato

This book
belongs to:

_____

Joaquim, since he was very young, loved to look at the night sky. He was completely amazed by the immensity of the universe and wanted to learn as much as he could about astronomy, which is what humanity knows about it.

How they shine!

1

When he was only 5 years old, his father took him to see the stars through a telescope. It was just an amateur instrument, but the little boy had an infinite curiosity.

On that day, Joaquim was fascinated by the beauty of the sky and by the stories his father told him about the constellations. He learned how ancient peoples navigated by the stars, the great voyages of exploration, and the stories of great astronomers.

3

Since then, every night before going to bed, Joaquim would look at the sky through the window of his room.

4

He always dreamed, sometimes even during waking hours, of being able to travel to space and explore the planets and stars that he admired so much.

One day, when Joaquim was 7 years old, he asked his father, 'Dad, what is the name of that star that shines brighter than the others in the sky?

The father explained that the bright spot was not a star, but rather the planet Venus. _My son, this bright point, although often mistaken for a star, is the planet Venus. Due to its thick and dense atmosphere, sunlight is strongly reflected, causing that intense brightness and many call it the morning star.

Joaquim was intrigued by the constant brightness of the planet Venus and the fact that stars appeared to twinkle in the sky. Once again, with his insatiable curiosity, he asked his father how this could happen.

And he learned that, because they are so far away, stars appear to be just bright points, and their light, when entering Earth's atmosphere, undergoes various reflections and deviations, causing them to twinkle, a phenomenon known as scintillation.

9

However, the planet Venus, being much closer, behaves like a larger body, and the sunlight reflected on its surface enters Earth's atmosphere without the reflections and deviations being noticed.

From that moment on, Joaquim began to read everything he could about the planet and space in general. He was amazed when he discovered that a day on Venus lasts about 243 Earth days, while a Venusian year is equivalent to approximately 225 Earth days.

So a Venutian could celebrate their birthday twice in the same day. How cool!

This difference in duration between the Venusian day and year is a result of the planet's retrograde rotation, meaning the rotational movement is in the opposite direction to the translational movement. This phenomenon is unique in the Solar System and is likely due to some collision with an asteroid.

'm upside down spinning.

The father of the curious young man always encouraged him to learn more and more about science and astronomy. He bought books, magazines, and educational videos for his son. Joaquim spent hours reading and watching everything he could find.

Never forget that curiosity is as important as intelligence. And always have discipline.

With the passing of the years, Joaquim became an astronaut. On his last mission, he commanded a team that visited the Moon, and now he would lead the first interplanetary trip that would take humans to Venus.

During the journey, the spacecraft commanded by Joaquim and his team lost contact with Earth when they passed through the dark side of the moon.

Concerned about what might have happened, Joaquim and his colleagues investigated the cause of the communication failure.

Sophia, one of the astronauts, suggested that an extraterrestrial spacecraft could have blocked the rocket's signal.

However, Joaquim remembered his father's teachings about Occam's Razor, which says that the simplest explanation is usually the correct one. Thinking about an extraterrestrial spacecraft seemed absurd at the moment, and Mike, one of the astronauts on the team, suggested that the problem must be in the transmitter circuit, the simplest and most reasonable explanation.

My little one, when in doubt, always stick with science!

In a few minutes, they discovered that the problem was just a blown fuse and managed to fix it, restoring communication with Earth. With the problem solved, they were able to proceed calmly towards Venus.

When the spacecraft arrived on Venus, they were surprised by the extreme conditions of the planet.

The temperature on the surface was about 462 degrees Celsius, which is enough to melt lead. Additionally, the atmospheric pressure is about 90 times higher than that on Earth, which is equivalent to being 900 meters below water.

Venus's atmosphere is primarily composed of carbon dioxide and sulfuric acid, rendering the planet inhospitable to life as we know it on Earth.

Even with these extremely harsh conditions, they managed to explore the upper atmosphere of the planet and collect many important data points.

Despite the team's desire to land on Venus, the conditions prevented any attempt. Thus, it was decided to send a robot to gather data and information about the planet. The robot endured for a few hours but ended up being destroyed. Nevertheless, the collected data proved to be extremely valuable for the astronaut team.

The high temperature on Venus is a result of an extremely intense greenhouse effect, which is caused by the presence of gases like carbon dioxide in its atmosphere.

This situation made the entire team of astronauts reflect on the importance of protecting Earth's atmosphere and controlling the emission of gases that contribute to global warming and the greenhouse effect on our planet. We still have much to learn about our own home, and space exploration can help us better understand our place in the universe.

Throughout the adventure, Joaquim couldn't help but wonder if there were extraterrestrial beings on Venus. He searched every corner of the planet for signs of life, but found nothing.

At the end of the mission, Joaquim and his team returned home safely, carrying a wealth of important data and new discoveries about the planet Venus. However, in his mind, Joaquim still wondered if he would ever uncover the answer to the big question: Are there extraterrestrial beings in the universe?

The end.